I0186685

The

Chronicles

Of

Iris

The

Chronicles

Of

Iris

...futuristic predictions

Cosmic Predictions
Volume 2
Copyright 2020

To

those

who

believe

Cosmic Predictions
Volume #2
Table of Contents

Cosmic Predictions
Volume #2
Table of Contents

Cosmic Predictions
Volume #2
Table of Contents

The Chronicles of Iris

...futuristic Predictions

Cosmic Predictions
Volume 2

Disclaimer

Chronicled Futuristic Predictions are
intuitive opinions made by the Author.
This book is protected under the
First Amendment of United States
Constitution.
For Entertainment Purposes Only
ReadersViewer discretion advised

Anatomy of a Chronicled Prediction

Origin of the Prediction:

The Chronicles of Iris

...futuristic predictions

Chronicle number: (01012018-A)
(month, day, year, letter)

Date of Chronicle Prediction: January 1, 2018

Prediction Number: Prediction 1

Prediction: "Ocean temperatures continue to rise

The Chronicles of Iris

...futuristic predictions

Cosmic Predictions
Volume 2

The Chronicles of Iris
…futuristic predictions

Chronicle (11172018-A)

November 17, 2018

Prediction 1

"Carbon dioxide detected between the Earth and the Moon"

The Chronicles of Iris

...futuristic predictions

Chronicle (11172018-B)

November 17, 2018

Prediction 2

"Space Antennae

Deepstep

is constructed"

The Chronicles of Iris

…futuristic predictions

Chronicle (11182018-C)

November 18, 2018

Prediction 3

"The interior

of

Venus is habitable"

The Chronicles of Iris

...futuristic predictions

Chronicle (11192018-D)

November 19, 2018

Prediction 4

"New Space energy

source

is identified "

The Chronicles of Iris
...futuristic predictions

Chronicle (11192018-E)

November 19, 2018

Prediction 5

"The interior
of Saturn's Moon
Enceladus is habitable"

The Chronicles of Iris

...futuristic predictions

Chronicle (11182018-F)

November 18, 2018

Prediction 6

"The temperature of the Sun's surface is growing hotter"

The Chronicles of Iris
...futuristic predictions

Chronicle (11182018-G)

November 18, 2018

Prediction 7

"Luna, Earth's moon assist in stabilizing Earth's atmosphere"

The Chronicles of Iris

...futuristic predictions

Chronicle (11192018-H)

November 19, 2018

Prediction 8

"Earth's moon, Luna could alter the Earth's tilt"

The Chronicles of Iris

...futuristic predictions

Chronicle (11192018-J)

November 19, 2018

Prediction 9

"Earth's moon,
Luna's interior
is hollow"

The Chronicles of Iris

...futuristic predictions

Chronicle (11192018-J

November 19, 2018

Prediction 10

"Evidence of life sustaining
elements uncovered inside
Earth's moon, Luna"

The Chronicles of Iris

...futuristic predictions

Chronicle (11192018-K)

November 19, 2018

Prediction 11

"*Information validating a platform for Life exists on the interior of Earth's Moon, Luna*"

The Chronicles of Iris

...futuristic predictions

Chronicle (11192018-L)

November 19, 2018

Prediction 12

"Earth's Moon, Luna's surface composition reveals a new element"

The Chronicles of Iris

...futuristic predictions

Chronicle (11202018-M)

November 20, 2018

Prediction 13

"Earth's Moon, Luna unfolds her secrets, slowly"

The Chronicles of Iris

...futuristic predictions

Chronicle (11202018-N)

November 20, 2018

Prediction 14

"Earth's moon, Luna reveals new data verifying water existing on the interior"

The Chronicles of Iris

...futuristic predictions

Chronicle (11202018-O)

November 20, 2018

Prediction 15

"Earth's moon, Luna, exposes its dark side revealing hidden civilizations"

The Chronicles of Iris
...futuristic predictions

Chronicle (11202018-P)

November 20, 2018

Prediction 16

"Elevated levels of oxygen detected in Luna's atmosphere Earth's Moon"

The Chronicles of Iris

...futuristic predictions

Chronicle (11202018-Q)

November 20, 2018

Prediction 17

"Luna's interior temperature is colder then it's surface"

The Chronicles of Iris

...futuristic predictions

Chronicle (11202018-R)

November 20, 2018

Prediction 18

"Venus' axial tilt shifts multiple degrees"

The Chronicles of Iris

...futuristic predictions

Chronicle (11202018-S)

November 20, 2018

Prediction 19

"Venus' spin shifts multiple degrees West"

The Chronicles of Iris

...futuristic predictions

Chronicle (11202018-T)

November 20, 2018

Prediction 20

"Venus' core is 2500-3000 miles in radius"

The Chronicles of Iris

...futuristic predictions

Chronicle (11202018-U)

November 20, 2018

Prediction 21

"Venus' highest mountain is hidden under the clouds 3500-6500 feet high"

The Chronicles of Iris

...futuristic predictions

Chronicle (11202018-V)

November 20, 2018

Prediction 22

"Venus has an impact crater five+ miles wide"

The Chronicles of Iris

...futuristic predictions

Chronicle (11202018-W)

November 20, 2018

Prediction 23

Venus' atmosphere
of Carbon dioxide
is utilized for fuel"

The Chronicles of Iris

...futuristic predictions

Chronicle (11202018-X)

November 20, 2018

Prediction 24

"Venus' atmosphere heat exceeds 1000-1500 degrees Fahrenheit"

The Chronicles of Iris

...futuristic predictions

Chronicle (11202018-Y)

November 20, 2018

Prediction 25

"Venus' magnetic field reverses the axel's rotation"

The Chronicles of Iris

...futuristic predictions

Chronicle (11202018-Z)

November 20, 2018

Prediction 26

"The Milky Way is over 6 billion years old "

The Chronicles of Iris

...futuristic predictions

Chronicle (11202018-A1)

November 20, 2018

Prediction 27

"The composition of Venus' surface contains chemicals used for fuel"

The Chronicles of Iris

...futuristic predictions

Chronicle (11202018-B1)

November 20, 2018

Prediction 28

"Venus' rotation changes direction"

The Chronicles of Iris

...futuristic predictions

Chronicle (11202018-C1)

November 20, 2018

Prediction 29

"Venus atmosphere is cooler in different regions on the planet"

The Chronicles of Iris

...futuristic predictions

Chronicle (11202018-D1)

November 20, 2018

Prediction 30

"Venus

possesses

hidden Oceans"

The Chronicles of Iris

...futuristic predictions

Chronicle (11202018-E1)

November 20, 2018

Prediction 31

"A region of Venus is desert"

The Chronicles of Iris

...futuristic predictions

Chronicle (11202018-F1)

November 20, 2018

Prediction 32

"A region of Venus has rivers"

The Chronicles of Iris

...futuristic predictions

Chronicle (11202018-G1)

November 20, 2018

Prediction 33

"There is of a
mountain range
on Venus "

The Chronicles of Iris

...futuristic predictions

Chronicle (11202018-H1)

November 20, 2018

Prediction 34

"The surface of Venus

is over

500 million years old"

The Chronicles of Iris

...futuristic predictions

Chronicle (11202018-J1)

November 20, 2018

Prediction 35

"A Solar Day on Venus is longer than 243 Earth days"

The Chronicles of Iris

...futuristic predictions

Chronicle (11202018-J1)

November 20, 2018

Prediction 36

"A Solar Day on Mars is longer than 25 hours"

The Chronicles of Iris

…futuristic predictions

Chronicle (11202018-K1)

November 20, 2018

Prediction 37

"A Solar Day

on Jupiter

is longer

than 10 hours"

The Chronicles of Iris

...futuristic predictions

Chronicle (11202018-L1)

November 20, 2018

Prediction 38

"A Solar Day
on Saturn
is longer
than 11 hours"

The Chronicles of Iris

...futuristic predictions

Chronicle (11202018-M1)

November 20, 2018

Prediction 39

"A Solar Day
on Uranus
is longer
than 17 hours"

The Chronicles of Iris

...futuristic predictions

Chronicle (11202018-N1)

November 20, 2018

Prediction 40

"A Solar day
on Neptune
is longer
than 16 hours"

The Chronicles of Iris

...futuristic predictions

Chronicle (11202018-01)

November 20, 2018

Prediction 41

"A Solar day
on Mercury is longer
than 1,408 hours"

The Chronicles of Iris

...futuristic predictions

Chronicle (11202018-P1)

November 20, 2018

Prediction 42

"There is a miscalculation in measurement for the Sidereal Day affecting many of the planets in the Milky Way"

The Chronicles of Iris

...futuristic predictions

Chronicle (11202018-Q1)

November 20, 2018

Prediction 43

"New measurement calculation
are adopted for the Solar Day
and Sideral Day for many of the
planets in the Milky Way"

The Chronicles of Iris

...futuristic predictions

Chronicle (11202018-R1)

November 20, 2018

Prediction 44

"A Sun

without orbiting

planets, discovered"

The Chronicles of Iris

...futuristic predictions

Chronicle (11202018-S1)

November 20, 2018

Prediction 45

"A portion of Venus' heat escapes the Cloud-covered atmosphere"

The Chronicles of Iris

...futuristic predictions

Chronicle (11202018-T1)

November 20, 2018

Prediction 46

"Venus slows down

appearing to turn clockwise

in retrograde rotation

1-3 times in a

Venusian year"

The Chronicles of Iris

...futuristic predictions

Chronicle (11202018-U1)

November 20, 2018

Prediction 47

"Venus' equator rotates

faster than

4.05 mph"

The Chronicles of Iris

...futuristic predictions

Chronicle (11202018-V1)

November 20, 2018

Prediction 48

"Venus shifts

the

Rotational speed"

The Chronicles of Iris

…futuristic predictions

Chronicle (11202018-W1)

November 20, 2018

Prediction 49

"The Sun rises in the East and sets in the West on Venus when the retrograde rotation changes its direction"

The Chronicles of Iris

...futuristic predictions

Chronicle (11202018-X1)

November 20, 2018

Prediction 50

"Venus is stabilized by Earth built Satellite / Moon"

Purchase my other chronicled

prediction books from Amazon

books & Kindle

Watch & subscribe to
The Chronicles of Iris Podcast

The Chronicles of Iris..
YouTube

www.thechroniclesofiris.com

www.thechroniclesofiris.com

www.ingramcontent.com/pod-product-compliance
Lightning Source LLC
Chambersburg PA
CBHW051710090426
42736CB00013B/2635